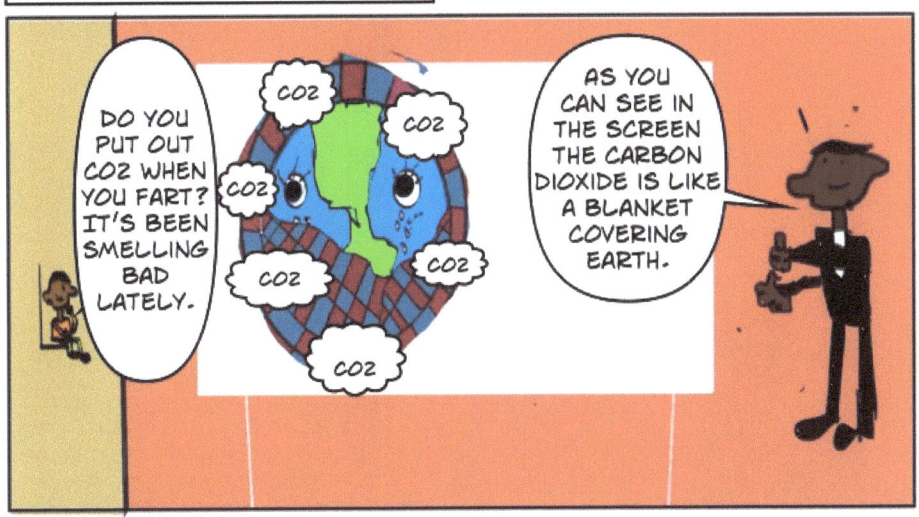

CFC'S ARE GASES THAT MAKE THE FRIDGE COOLER BUT MAKE THE PLANET HOTTER.

SOMETIMES THE FRIDGE LEAKS AND THE CFC'S ESCAPE.

THE CFC MOLECULE GETS TO THE ATMOSPHERE

THE CFC MOLECULE STARTS CONTRIBUTING TO THE HEAT BLANKET.

CLIMATE CHANGE MAKES EARTH HEAT UP SO MUCH IT CAUSES WILDFIRES.

AFTER THE FIRE...

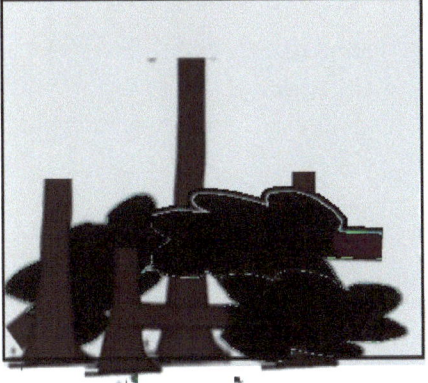

WE NEED TO SAVE TREES BECAUSE TREES TAKE IN CARBON DIOXIDE AND GIVE OUT OXYGEN.

CLIMATE CHANGE MAKES EARTH SO HOT IT CAUSES GLACIERS TO MELT INTO THE OCEAN WHICH MAKES THE SEA LEVEL RISE THEN CAUSES FLOODS.

GLACIERS MELTING IN ALASKA -
YOU'RE ABOUT TO FALL INTO THE WATER
NO I'M NOT
I'M ON A TOTALLY STURDY LAND.

MEANWHILE, IN FLORIDA...
HUH, I WONDER WHY IT'S FLOODING

CLIMATE CHANGE CAN ALSO DESTROY CORAL REEFS.

THE WARMTH FROM THE HEAT TRAPPING GASES SURROUNDING EARTH MAKES THE REEF WARMER.

THEN THE CORAL REEF DIES.

CLIMATE CHANGE MAKES EARTH SO HOT IT CAUSES DROUGHTS AND CAN EVEN CREATE DESERTS.

THE HEAT CAUSES THE LAND TO GET DRY AND THE LAKE WILL EVAPORATE

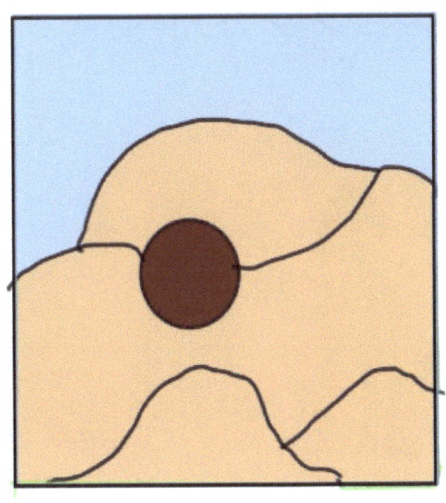

PART 2

ACTUALLY THAT IS A WAY TO HELP CLIMATE CHANGE. BUT WE DON'T NEED TO STOP FARTING, THE COWS DO!

WHEN COWS FART OR BURP IT CREATES METHANE THAT GOES INTO THE ATMOSPHERE AND MAKES IT THICKER.

IF HUMANS EAT LESS COWS WE CAN REDUCE THE NUMBER OF COWS WE HAVE IN FARMS. IF WE HAVE LESS COWS IN FARMS THERE WILL BE LESS METHANE IN OUR ATMOSPHERE.

COULD I GET 1 ELECTRIC CAR PLEASE.
SURE

DUMP
YOU'VE GOT TO BE KIDDING ME.

ANOTHER SOLUTION TO CLIMATE CHANGE IS TO ADD SOLAR PANELS ON YOUR HOUSE.

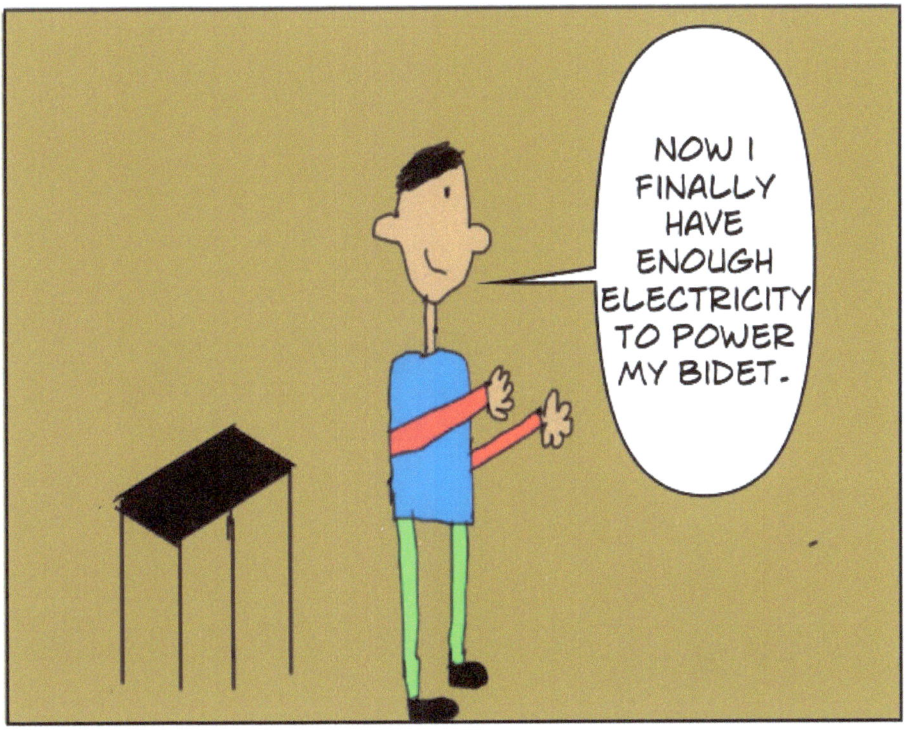
NOW I FINALLY HAVE ENOUGH ELECTRICITY TO POWER MY BIDET.

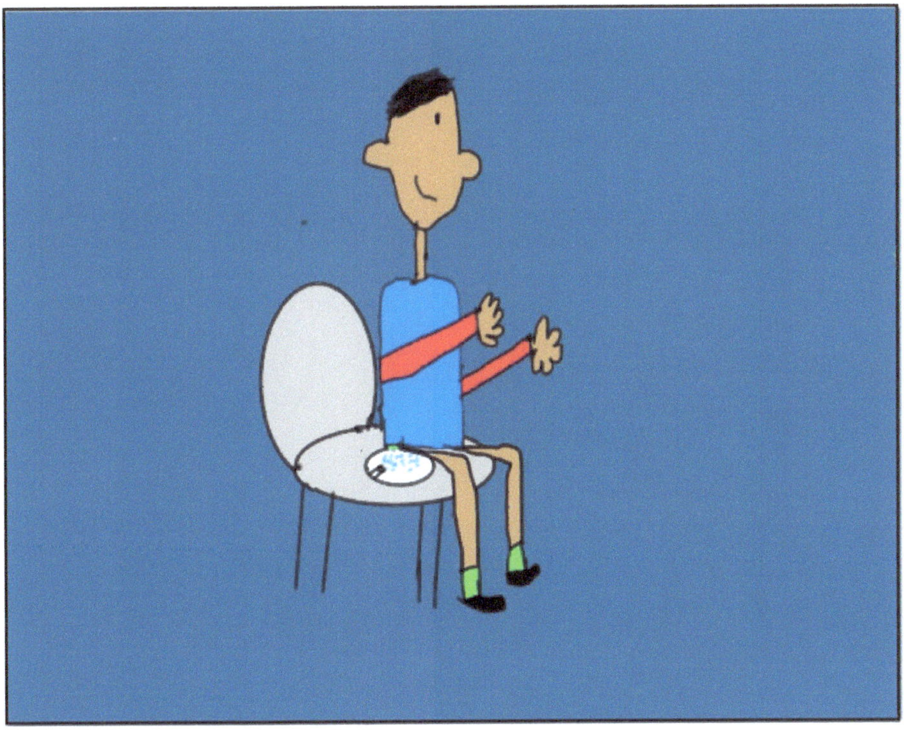

PART 3

YOU CAN HELP SOLVE CLIMATE CHANGE BY INFORMING OTHERS.

AYAN IS 12 YEARS OLD. HE LIVES IN SAN DIEGO, CALIFORNIA IN PACIFIC BEACH. HE HAS SO MUCH FUN PLAYING TENNIS WITH HIS DAD. AYAN ALSO PLAYS AN INSTRUMENT CALLED SITAR WITH HIS MOM. HE ALSO LIKES TO PLAY SPORTS WITH HIS YOUNGER SISTER. HE LOVES TO READ. HE ESPECIALLY LIKES TO READ COMIC BOOKS. HE GOT INSPIRED TO WRITE THIS BOOK BECAUSE HE WANTS TO EDUCATE KIDS ABOUT CLIMATE CHANGE AND HE ALSO HAD A GREAT TIME WRITING AND DRAWING THE BOOK.

www.ingramcontent.com/pod-product-compliance
Lightning Source LLC
Chambersburg PA
CBHW050912290526
45792CB00002B/790